Cute and Cuddly: Baby Animals

LAMBS

By Katie Kawa

Gareth Stevens Publishing

Please visit our website, www.garethstevens.com. For a free color catalog of all our high-quality books, call toll free 1-800-542-2595 or fax 1-877-542-2596.

Library of Congress Cataloging-in-Publication Data

Kawa, Katie.
Lambs / Katie Kawa.
p. cm. — (Cute and cuddly–baby animals)
ISBN 978-1-4339-5547-1 (pbk.)
ISBN 978-1-4339-5549-5 (6-pack)
ISBN 978-1-4339-5546-4 (library binding)
1. Lambs—Juvenile literature. I. Title.
SF376.5.K39 2011
636.3'07—dc22

2010052585

First Edition

Published in 2012 by
Gareth Stevens Publishing
111 East 14th Street, Suite 349
New York, NY 10003

Editor: Katie Kawa
Designer: Andrea Davison-Bartolotta

Photo credits: Cover, pp. 1, 7, 9, 13, 15, 19, 23, 24 (mountain) Shutterstock.com; pp. 5, 11, 17, 24 (hay) iStockphoto.com; pp. 21, 24 (wool) David De Lossy/Photodisc/Thinkstock.

Printed in the United States of America

CPSIA compliance information: Batch #CS11GS: For further information contact Gareth Stevens, New York, New York at 1-800-542-2595.

Contents

A lamb is a baby sheep.

Lambs walk soon after they are born.

Lambs are born in the spring.

A lamb makes noise.
The noise is called
a bleat.

A lamb stays close to its mother. A mother lamb is called a ewe.

Lambs drink milk at first.

Soon, lambs eat grass. They eat hay in the winter.

Lambs have soft hair.
This is called wool.

People use wool to make clothes!

Lambs live in the mountains. They live on farms too.

Words to Know

hay

mountains

wool

Index